TYPEWRITER

TYPEWRITER

THE HISTORY * THE MACHINES * THE WRITERS

BY TONY ALLAN
CONSULTANT: RICHARD POLT

SHELTER HARBOR PRESS
NEW YORK

A sultry secretary
from the 1920s
perches beside her
typewriter.

ISBN 978-1-62795-034-3

Created by Toucan Books Ltd.
Text: Tony Allan
Consultant: Richard Polt
Editor: Dorothy Stannard
Designer: Leah Germann
Picture research: Christine Vincent
Proofreader: Marion Dent
Index: Marie Lorimer

SHELTER HARBOR PRESS

12 11 10 9 8 7 6 5 4 3 2 1

Printed in China
First printing, 2015

CONTENTS

Marilyn Monroe
shows some leg in
this 1949 ad.

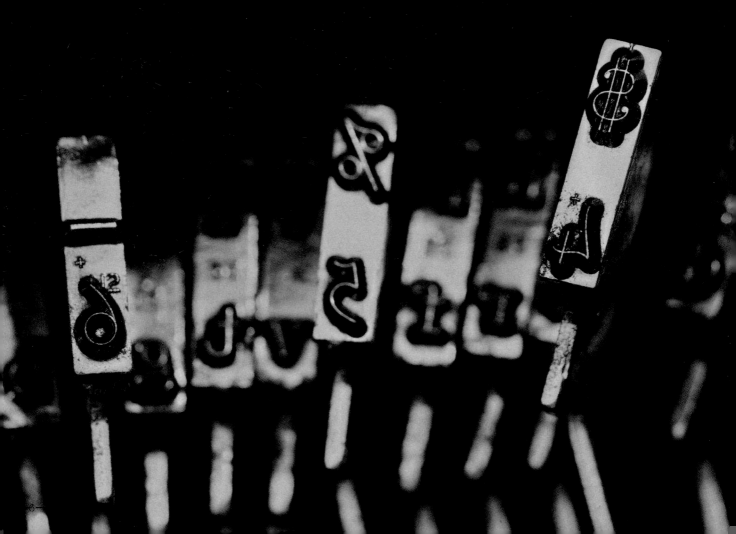

FOREWORD

I've been a typewriter repairman for 55 years and
although I've introduced plenty of people to typewriters,
this is the first time I've ever been asked to introduce
a book. Leafing through the following pages, I've seen
old friends like the 100-year-old Underwood No. 5,
the 1923 Corona portable, and the 1961 IBM Selectric. And
I've learned more about typewriter history, from early
inventors to modern-day aficionados.

    I'm glad that typewriters are back in vogue. It's
a pleasure to see kids coming into the shop saying they
want to write and getting their first typewriters. I hope
this book will inspire you to get typing.

PAUL SCHWEITZER,
Gramercy Typewriter Company, Est. 1932

CHAPTER 1:
* THE BIRTH OF TYPING *

As many as 52 different people are credited with
inventing the typewriter, though their designs had
little in common with the modern machine. These curious
contraptions included coffin-shaped boxes, piston-
studded domes, and piano-like instruments that look
more suited to composing a sonata than a sonnet.
Inventors faced an array of challenges--such as getting
the words to travel across the page and the need to
type the entire alphabet, in capitals and lowercase
(an early demo models could type just one letter--"w").
By 1878, Remington had most of the elements in place;
in 1901, Underwood made the first modern typewriter.

MERCEDES

Schreibmaschine

REKLAMEVERLAG ERNST MARX BERLIN W 8

Mercedes·Bureau·Maschinen·G·m·b·H·Berlin·W·30

IN THE BEGINNING

People imagined a writing machine long before there was one, much as we might dream of flying cars today. Inspired by the success of the printing press, invented in 1450, the educated classes desired something clearer, cleaner, and, above all, quicker than writing with a quill pen. They also thought that print would give their words the validity and permanence of books.

Ideas for a writing machine floated around the 18th-century zeitgeist. In England, Queen Anne granted a patent for "an artificial machine or method for the impressing or transcribing of letters" to the engineer Henry Mill as early as 1714 (it isn't known how far he got with it). A desire to help the blind and deaf communicate set some inventors tinkering.

PAPER MOUNTAIN

The pace of invention accelerated in the 19th century. The industrial revolution generated a mountain of inventories, contracts, and accounts that had to be copied by an army of clerks——a breed embodied by Bob Cratchit, who toils away in a "sort of tank" in Charles Dickens's A Christmas Carol.

Inventors raced to design and make a "typographer" that would cut costs and save time. They experimented with hooks and levers, pistons and plungers. In Massachusetts in 1843, Charles Thurber invented the Patent Printer, a typewheel with a carriage to transport the paper. It was a big step forward but slow. The magazine Scientific American described it as "perfectly efficient except to the element of time."

"They cannot fool FATHER TIME,
No matter how they fool their time away."

Before typewriters took over, clerks spent 60 hours a week handcopying documents. In this advertisement for Bar-Lock typewriters, developed in New York in the 1880s, Father Time himself blows the whistle on this wasteful practice.

MIO CARO AMICO

A LABOR OF LOVE

Necessity was the mother of invention. In 19th-century
Italy, the nobleman Pellegrino Turri longed for
intimate letters from his lover, Countess Carolina
Fantoni da Fivizzano*, but the young noblewoman's
sight was fading and she could no longer see to write.
Inspired by pity, but also by passion and desire, in
1808 Turri invented a typewriter and a kind of carbon
paper ("black paper") to give his mistress the gift
of intimate communication without the assistance of
her maid. Sadly, we have no record of the engineering
specifications of Turri's typewriter, but 16 of the
countess's typed (all in capitals) billets-doux
survive. They begin "MIO CARO AMICO ..." (My darling
friend). It is said that when the Countess ran out of
carbon paper, she wrote: "I am desperate."

* Carey Wallace's 2010 novel The Blind
Contessa's New Machine imagines the affair
between Carolina and Turri.

MACHINES FOR THE BLIND

Turri wasn't the only inventor motivated by a desire to help the blind. Pietro Conti's Tachigrafo, invented in France in 1823, was designed for "even those with poor sight in mind," as was Charles Thurber's chirographer of 1845. The 1862 prospectus of Eastman Business College, in Poughkeepsie, New York, recommended Thurber's chirographer——a mechanically operated pen——to partially sighted students as well as to anyone with cramp or trying to write on a moving train. Even the first machine to be called a "Type Writer," the Sholes & Glidden of 1874, was publicized as a "great boon to the blind," especially following the invention of dictating machines in the 1880s and the introduction of audio-typing.

* * * * * * * * * *
The Braille typewriter, invented by Frank Hall in 1892, revolutionized communication for the blind. On meeting Hall, the deaf-blind activist Helen Keller (1880—1968), shown above, hugged and kissed him in gratitude.
* * * * * * * * * *

THE PINCUSHION

Popularly known as the
"pincushion machine,"
the Hansen Writing Ball
was the world's first
commercially produced
typewriter. Developed by
Rasmus Malling-Hansen,
the principal of the Royal
Institute for Deaf Mutes
in Copenhagen in 1865,
it featured 52 pistons
on a brass hemisphere
with vowels on the left
and consonants on the
right--the arrangement
found to work the
fastest. When pressed, the
pistons hit the paper via
carbonized paper or an
inked ribbon.

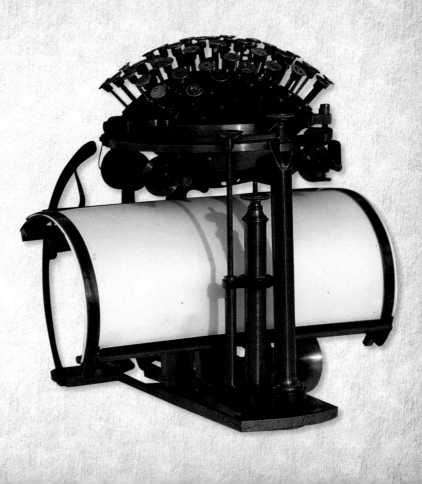

THE WRITER

The German philosopher Friedrich Nietzsche (1844—1900) owned a Hansen Writing Ball. Like the blind Contessa, who inspired Turri's machine (see p.12), Nietzsche had begun to lose his sight and bought the Writing Ball so that he could continue writing. But Nietzsche was bemused by his typewriter; he found it difficult to operate and complained that it influenced his thoughts. He was one of the first of many writers to suggest there could be a metaphysical connection between the means of writing and creativity.

"The writing ball is a thing like me:

Of iron, yet twisted easily.

Patience and tact must be had in abundance,

As well as fine little fingers, to use us."

* * * * * * * * * *
Nietzsche found his Hansen Writing Ball quirky. He wrote a wistful little poem in which he expressed an ironic sense of identification with it.
* * * * * * * * * *

THE EMERGING MARKET

The first U.S. patent was taken out on a writing
machine--called the Typographer and designed
by William Burt--as early as 1829. However,
another 44 years passed before typewriters
successfully went into production with the
Sholes & Glidden machine of 1873. Built in the
Remington sewing machine factory in Ilion, New
York, this was a refinement of a "Type-Writer
Machine" patented in 1868 by amateur inventors
Christopher Latham Sholes and Carlos Glidden and
a printer called Samuel Soule.

The first Sholes & Glidden typewriter left
the factory in 1874. It was hoped that businesses
as well as lawyers, writers, and the clergy
would buy the new machine. Initially, it sold
poorly--only 100 or so were bought during the
first year. A remodeled version aroused little
attention when it was presented at the Centennial
Exhibition in Philadelphia in 1876, the first
major World's Fair to be held in the U.S.

The 1868 patent for
the Type-Writer
Machine shows
how different
this forerunner
looked from the
Sholes & Glidden.

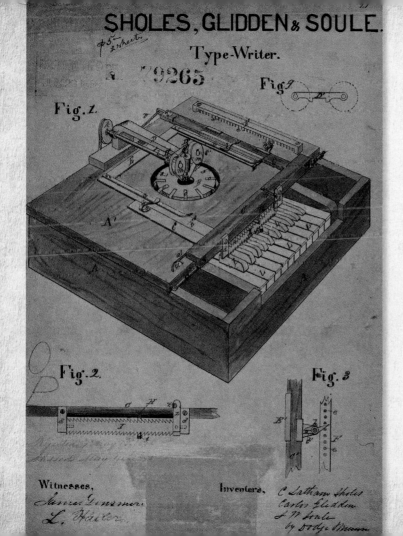

SHOLES, GLIDDEN & SOULE.

Type-Writer.

№ 79265

Fig. 1.

Fig. 9.

Fig. 2.

Fig. 3.

Witnesses,

Inventors,

C. Latham Sholes
Carlos Glidden
S. W. Soule
by Dodge & Mason

The Sholes & Glidden had four rows of capitals arranged in the QWERTY formation, though QWERTY wasn't included in the patent until the 1878 model.

THE MUSEUM OF LOST TYPEWRITERS

The early history of typewriting is littered with machines that left a name and sometimes a patent description but no model. Some of these casualties of history are listed below:

PIETRO CONTI'S TACHIGRAFO (1823)

This was the first machine to have upward-striking typebars. Designed to help those with poor sight, it was patented in France and tested by members of the Académie Française.

HENRY MILL'S WRITING MACHINE (1714)

A patent describes this as "an artificial machine or method for the impressing or transcribing of letters singly or progressively, one after another, as in writing."

WILLIAM BURT'S TYPOGRAPHER (1829)

This was the first writing machine to be patented in the U.S. A fire at the Patent Office destroyed the prototype, but a replica was made for the Chicago World's Fair of 1893.

XAVIER PROGIN'S
TYPOGRAPHIC PEN (1833)

The typebars of this design struck downward, allowing operators to see what they were typing. The bars were held in a circular basket and operated by levers.

CHARLES THURBER'S
CHIROGRAPHER (1845)

After inventing the Patent Printer in 1843, Thurber, from Worcester, Massachusetts, developed the Chirographer—a machine that operated a pen by means of piano keys.

ALEXANDER BAIN'S
PRINTING TELEGRAPH (1841)

The patent for this British device, designed to print telegrams, featured the ancestor of the typewriter ribbon—rubbed with oil, lamp black, and turpentine.

GIUSEPPE RAVIZZA'S
HARPSICHORD-WRITER (1855)

An Italian inventor, Ravizza produced 17 models over 55 years and introduced novelties such as a bell to signal line endings. His designs had keys similar to those on a piano.

TYPEWRITERS TAKE OFF

By the 1880s, typewriters were flying
out of the stores. Yet the early machines
had limitations. The original Sholes &
Glidden models, for example, could print
only capital letters. Furthermore, the
typebars were arranged to strike upward
on to the bottom of the platen (the rubber-
covered roller that holds the paper):
Operators had to lift the carriage to see
what they had typed.

 Things improved when Remington
introduced the shift mechanism, making it
possible to type both capital and lowercase
letters without the need for a seven-row
keyboard. The new device was incorporated
in the Remington No. 2 model of 1878. In
order for typists to view their work as
they went along, machines were redesigned
so that the typebars struck the top or
front of the platen, rather than the bottom.

Trained operators
such as this one,
photographed in
1892, were much
in demand.

NEW DEVELOPMENTS
Inventors kept working but not all their
innovations were successful. The nonstop
clatter of the new machines spurred the
production of "noiseless" models from 1910,
and there was even a Noiseless Typewriter
Company in Middletown, Connecticut.
Most of these machines failed to live
up to their name. Electric typewriters
appeared surprisingly soon, including the
Blickensderfer of 1902, made in Stamford,
Connecticut. For all its potential, however,
the model was ahead of its time; only three
of them are known to survive.

The Cahill electric
typewriter rivaled
the Blickensderfer.
It was light
and capable of
great speed.

Yost targets the French market with this 1928 advertisement designed by J. Stall.

IT PAYS TO ADVERTISE

By the early 1900s, typewriters satisfied all their purchasers' needs. They were easy to use, and fast, at least for people who could touch-type. The result was a massive increase in the popularity of the machines, and rival manufacturers fought for a share of the market. To boost sales in the U.S. and Europe, the big players turned to advertising. Remington hired Leonetto Cappiello, an Italian-born, Paris-based designer, to promote its wares, while the U.S. firm Yost commissioned exciting European talents such as French poster artist J. Stall and Hungarian artist Mihály Biró to spread the word about its "visible" models. Introduced in 1908, these were the first of the Yost machines to allow operators to view the words as they typed.

TRIUMPH

Triumph-Werke Nürnberg A.G.

German manufacturer Triumph-Adler began making typewriters in 1898. High spec design and appealing advertising made them a worldwide hit.

Bar-Lock extols the qualities of one of its early "visible" models.

THE FIRST MODERN TYPEWRITER

The Underwood No. 5, launched in 1901, was the first
truly modern typewriter. Designed by Franz Xavier
Wagner, a former Remington employee, it consolidated
and advanced design improvements made in the
four preceding Underwood machines, including the
introduction of visible printing. John T. Underwood,
the founder of the Underwood Typewriter Company,
had past links with Remington, having supplied the
company with typewriter ribbon when he ran his family
stationery business. When Remington began making
its own ribbon, John retaliated by starting his own
typewriter company. The No. 5 machine, with its light
touch and integral tabulator, finally got the formula
right: The model sold in the millions around the world
and remained in production with few improvements
for 30 years. Underwood became the world's leading
typewriter manufacturer.

>> At its peak, the Underwood factory in Connecticut rolled out one machine per minute.

>> The No. 5 model introduced a ribbon selector, allowing the operator to switch between red and black ink.

>> Underwood produced a giant typewriter for the Panama—Pacific International Exposition of 1915. It weighed 14 tons, and 16 "typists" were hired to sit on the keys. It was refurbished for the 1939 World's Fair.

>> Olivetti bought a controlling share of Underwood in 1959.

TYPEWRITER MUSEUMS

Many museums have collections celebrating the typewriter's heyday when its constant clatter filled offices all over the world.

 MILWAUKEE PUBLIC MUSEUM
Milwaukee, Wisconsin
The Dietz Collection in the hometown of typewriter pioneer Christopher Latham Sholes has more than 1,000 machines dating from the 1870s to the 1980s.

2 **MUSEE DES ARTS ET METIERS**
Paris, France
Models in this industrial design museum include a Yost No. 10 and Chinese and Japanese typewriters.

 MUSEUM OF BUSINESS HISTORY AND TECHNOLOGY
Wilmington, Delaware
This outstanding collection features an 1890 Smith Premier No. 1, a 1904 Mignon (shown left), and a Noiseless model from 1912.

 MUSEUM OF PRINTING HISTORY
Houston, Texas
A good section on typewriters sits within a wider story of mechanical printing.

 NATIONAL MUSEUM OF AMERICAN HISTORY
Washington, D.C.
Part of the Smithsonian Institution, this museum features the Sholes, Glidden & Soule patent model, the Yost No. 1, the Corona Silent, and the Keaton Music Typewriter, which printed musical notes.

 NATIONAL MUSEUM OF SCOTLAND
Edinburgh, Scotland
An early Sholes & Glidden from around 1875 and an Olivetti Valentine portable from the 1960s feature in a small collection of classic models.

 PETER MITTERHOFER TYPEWRITER MUSEUM
Parcines, Italy
Named for a typewriter pioneer from Parcines (formerly in Austria), this museum has more than 1,800 models. Among the 120 models on display are an Edison-Mimeograph Typewriter dating from 1894 and a Blickensderfer Oriental (1908) for typing Hebrew.

 POLYTECHNICAL MUSEUM
Moscow, Russia
A rare Kosmopolit index machine, produced in Hamburg from 1888, and a Blickensderfer Electric, from 1902, are among the highlights.

 SCIENCE MUSEUM
London, England
More than 100 models include an 1875 Sholes & Glidden, an 1889 Maskelyne with an inking pad instead of a ribbon, and a Blickensderfer.

TYPEWRITER TIMELINE

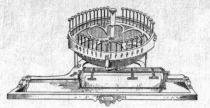

1808
Pellegrino Turri makes a machine for the use of his blind lover. Although the machine has not survived, several of her letters have--all in capital letters.

1843
Charles Thurber from Worcester, Massachusetts, invents a writing machine with a moving platen.

1874
The Sholes & Glidden, later known as the Remington No. 1, hits the market. Four years later, a refinement, the Remington No. 2, introduces the shift key.

1714
Henry Mill patents "an artificial machine or method for impressing or transcribing of letters ...one after another, as in writing." No designs or models of the machine survive.

1823
The Italian-born inventor Pietro Conti demonstrates an apparatus "writing fast and clear enough for anyone, even those with poor sight" before members of the Académie Française in Paris.

1868
Amateur inventors Christopher Latham Sholes, Carlos Glidden, and Samuel SoulE develop the prototype of the Sholes & Glidden machine. The prototype looks like a wooden box with six piano-style keys, but within six years it is recognizably a typewriter with a QWERTY keyboard.

1884
The Hammond introduces a **swiveling type shuttle** with a separate hammer that **drives** the paper against the letter from behind.

1910
The Noiseless Typewriter is launched, employing a complex lever mechanism that allows for quieter (though by no means noiseless) typing.

1961
The IBM 72 Selectric typewriter introduces the golf ball typehead, featuring all 88 standard characters.

1883
Mark Twain's Life on the Mississippi is the first major literary work to be submitted for publication in typewritten form.

1893
The Blickensderfer No. 5, the first commercially successful portable typewriter, is unveiled at the Chicago World's Fair.

1920
Remington releases a portable model with four rows of keys, as on modern machines.

The world went
crazy for QWERTY.
The German
version, QWERTZ,
inspired this
dance class in
1930s' Berlin.

CHAPTER 2 :
* QWERTY QUIRKS AND OTHER TYPES *

The keyboard arrangement invented for the Sholes &
Glidden typewriter of 1874 was called QWERTY, after
the first six characters on the top row of letters.
Designed to prevent the typebars from clashing, QWERTY
was adopted by typewriter manufacturers across much
of the world. Such ubiquity oiled the labor market and
simplified the teaching of touch-typing. It could even
help detectives solve crimes. Did a faintly inked "p" in
a typed blackmail note, for example, point to a worn-
down typehead on a badly maintained Underwood or a
sweet-looking touch typist whose little finger could
only just reach the key?

10 THINGS YOU MIGHT NOT KNOW ABOUT QWERTY

>> 1. When devising QWERTY, Christopher Sholes asked his son-in-law to draw up a list of the most common two-letter combinations in English, then apparently designed a keyboard that separated the letters so that the typebars would not clash.

>> 2. Initially, the top row read QWE.TYIUOP, with a period after the E; in later models this was rationalized to QWERTYUIOP. Foreign adaptations included QWERTZUIOPÜ (Germany) and QWERTYUIOPÅ (Scandinavia).

>> 3. Sholes got Washington, D.C. court reporter James Ogilvie Clephane to test QWERTY. Clephane trashed at least three machines while doing so.

>> 4. Although other companies adopted QWERTY, they were not charged patent royalties—perhaps because Remington also had a growing chain of typing schools and wanted to consolidate the QWERTY system.

>> 5. Sholes had doubts about QWERTY, and experimented with different keyboards. The final one, patented in 1889, started XPMCHRT, and grouped all the vowels on the right-hand side of the middle row of keys.

>> 6. The original Sholes & Glidden keyboard did not feature the numbers 0 and 1. Operators were expected to type the capital letters O and I instead.

>> 7. Although the rationale behind QWERTY was to stop the typebars from clashing, one common letter pairing, "e" and "r," lie side by side on the keyboard while the typebars they control are separated by just one typebar.

>> 8. In 2013, researchers in Kyoto, Japan, upset QWERTY lore by asserting that it was devised to suit telegraph operators translating Morse code.

>> 9. More words can be typed using just the letters for the left hand than can using those for the right—TEN times more according to some reports.

>> 10. QWERTY's supremacy is being challenged. Techies are now looking to new keyboards such as KALQ, designed for thumb typing on smart devices.

RIVAL KEYBOARDS

Although the Sholes & Glidden machine used the four-row keyboard (three rows of letters and one of numbers and symbols) that is now standard, there were other arrangements in the early days, each of which claimed to be faster and better.

Double keyboards had lowercase letters at the bottom and capitals at the top. This seven-row layout made touch-typing impossible and the machine therefore slow. The brand Smith Premier persisted in using the double keyboard until 1921.

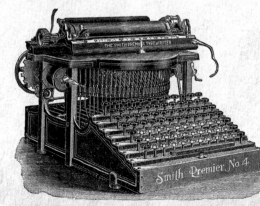

Smith Premier No. 4 with double keyboard

A three-row machine
made by Oliver

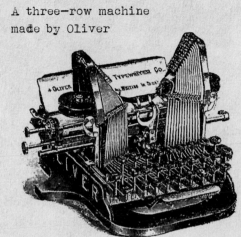

There were also three-row QWERTY
machines with a double shift that
allowed each key to type three
different characters. This setting
proved popular for portables, as
having fewer typebars reduced space
and weight. One German model, the
Helios, experimented with just two
rows of characters and a triple shift.

AN ONGOING QUEST

The semi-random nature of the
QWERTY keyboard encouraged the
invention of other layouts that
might increase typing speed. The
inventive Blickensderfer company
experimented with having a
DHIATENSOR sequence on the bottom
row, based on the most commonly
typed letters. In 1936, August
Dvorak, an American university
professor, and his brother-in-
law William Dealey patented the
Dvorak Simplified Keyboard. Its
layout grouped the five vowels
on the far left of the middle row
of letters and concentrated the
most used consonants in the right-
hand register. Although there is
no definitive proof of Dvorak's
superiority, all major modern
computer operating systems support
it as an option, allowing users to
switch from QWERTY if they wish.

* * * * * * * * * * * * * * * * * * *

Paper shelf

Platen or roller

Paper locating guide

Line space adjusting lever

Line space indicator

Carriage release

Platen clutch release

Platen turning knob

Line space lever

Margin indicator

Backspacer

Shift key

Paper finger

Paper release

Paper holder

Ribbon carrier

Position indicator

Margin scale

Typebars

Ribbon color selector

Shift lock

Shift key

Space bar

```
* * * * * * * * * * * * * *
```

OIL AND BRUSHUP
The ideal secretary in
the 1950s paid as much
attention to the upkeep
of her typewriter as she
did to her appearance.
The Typist's Treasury,
issued by England's
Imperial typewriter
company, spelled out
the routine: "Clean the
machine daily with
duster and long-handled
soft brush; oil carriage
rails and typebars
sparingly; wipe the type
itself with a stiff
brush after lengthy
spells of typing."

```
* * * * * * * * * * * * * *
```

RIBBONS GALORE

The only technical difficulty most typists faced was changing the ribbon--an inky but necessary task that could transform the quality of their work.

EVERYTHING FOR THE OFFICE

TELEPHONE WABASH 1807

```
* * * * * * * * * * * * *
```

TYPEWRITER TOOL KIT

For cleaning
* Q-tips
* Soft cloths/rags
* Alcohol
* Acetone
* Metal polish
* Fine automotive
  rubbing compound
* Automobile wax

Tools
* Precision screwdrivers
  (gunsmith
  screwdrivers)
* Needle-nosed pliers
* A 5/16th combination
  wrench
* A small brass hammer
* Air compressor (for
  blowing out cobwebs
  and dust)
* Replacement screws
  (if needed)

```
* * * * * * * * * * * * *
```

Dearest...

"A typewriter
has quite as much
individuality as a
man's handwriting."
-- Sherlock Holmes

CRIME AND PUNISHMENT

Over the years typewriters came to play a significant
role in the fight against crime, especially fraud
and blackmail. The idiosyncrasies of an individual
machine, such as the misalignment of a certain letter,
the pressure exerted on the keys by its operator,
or the subtle variations in the typeface of different
makes and models enabled detectives to trace the origin
of typed documents.

     As early as 1893, a judge trying Levy v. Rust in
a New Jersey courtroom described sitting down with
"an expert in typewriting" to examine some contested
documents under a magnifying glass. "I was very much
struck by his evidence," the justice concluded. "If you
compare the typewriting work, it contains precisely the
same peculiarities which are found in the typewriting
in these seven suspected papers." Strangely, the FBI
did not launch the Document Section of its crime lab to
follow up such clues until 1932.

NEWNES SIXPENNY COPYRIGHT NOVELS

ADVENTURES OF SHERLOCK HOLMES

BY
A. CONAN DOYLE

* *

<u>THE ULTIMATE SLEUTH</u>
Sherlock Holmes was ahead of the game when it came to solving a crime with a typewriter. In Arthur Conan Doyle's 1891 story "A Case of Identity," the third story in the <u>Adventures of Sherlock Holmes</u>, the detective proves that typewritten love letters sent to a young woman were written by her stepfather. The motive, Holmes concludes, was to stop her from finding a suitor who might threaten access to her inheritance.

* *

One of the most famous
typewriters in legal
history featured in a case
involving not murder or
fraud but spying. This was
the Alger Hiss trial, which
split U.S. opinion in the
early Cold War years. In
1948, Whittaker Chambers,
a former member of the
Communist Party, accused
Hiss (a State Department
official who had helped
set up the United Nations
after World War II) of
being a Communist in front
of the House Un-American
Activities Committee. Hiss
denied the charge and
started libel proceedings

against his accuser. In retaliation Chambers stated that he and Hiss had both spied for Stalin's Soviet Union in the 1930s. As evidence, he produced a file of 65 retyped State Department documents that he claimed had been copied by Hiss's wife (Hiss himself could not type). The machine used was a Woodstock model, No. 230099.

Hiss eventually went on trial for perjury rather than treason, as time had run out under the statute of limitations. Much of the case against Hiss involved linking the documents to his machine. Even after he was found guilty and sentenced to five years in jail, Hiss continued to maintain that the documents had been faked. To make the point, his defense team hired renowned typewriter expert Martin Tytell to spend two years building a replica of the Woodstock, so accurate in every detail that experts would claim that documents typed on it had come from the original machine. Their efforts were in vain, however. The judge at Hiss's appeal refused a retrial.

* * * * * * * * * * * * * * * *
REVERTING TO TYPE
One advantage typewriters have over computers is that they can't be hacked. This has led some organizations to reevaluate their worth. The German government was reputedly dusting down its old machines following Edward Snowden's revelations about U.S. spying on Chancellor Angela Merkel in 2014. Reportedly, Russia took similar steps in 2013.
* * * * * * * * * * * * * * * *

CHAPTER 3 :
* THE CULTURED MACHINE *

A beat-up typewriter stands on a table or desk,
lit by lamplight. A half-drained glass of Scotch
rests beside it, as smoke curls nonchalantly upward
from a cigarette set aside in an ashtray. The scene
immediately says "writer," of course, just as a blank
canvas on an easel says "painter."

Bookended between quill pen and computer, the
typewriter established a unique position as a cultural
icon. Something about the urgent mechanical drive of
typing and the insistent clatter of the keystrokes and
line-end bells reflected the spirit of a dynamic age
and helped define 20th-century literature.

"The machine has several
virtues. I believe it will
print faster than I can
write ... It piles an awful
stack of words on one page.
It don't muss things or
scatter ink blots around."
—Mark Twain (writing
to his brother about his
new typewriter)

LEAVING A MARK ON LITERATURE

The first significant literary work to be
delivered to the publisher in typewritten
form is generally held to be Mark Twain's Life
on the Mississippi, in 1883. (Twain himself
claimed the work was The Adventures of Tom
Sawyer, but scholars over the years reckon he
was mistaken.)

The great humorist was an early convert
to typewriters, which he first encountered in
a Boston store in 1874. A salesman assured him
that the machines could turn out 57 words a
minute, and when Twain disputed the claim (the
maximum speed for handwriting being about 30
words) a demonstrator was summoned to prove it.
Again and again she rattled out her 57 words,
and it was only later, when Twain had a chance
to examine the slips, that he realized she was
simply repeating a prepared script from memory.
Nevertheless, he was sufficiently impressed to
buy one of the newfangled contraptions.

Mark Twain
taught himself
the rudiments
of typing by
endlessly tapping
out the phrase "The
boy stood on the
burning deck" until
he could manage 12
words a minute.

* * * * * * * * * * *

Twain's first
typewriter was a
Sholes & Glidden
costing $125.

* * * * * * * * * * * *
"There is nothing
to writing. All
you do is sit down
at a typewriter
and bleed."

-- Ernest Hemingway
* * * * * * * * * * * *

Hemingway at his
typewriter in
1944, when he was
working as a war
correspondent.

THE WRITE STUFF

Hemingway's favorite typewriter was the Royal Quiet
Deluxe Typewriter, although he owned other Royals over
the years. Papa Hemingway wasn't the only writer with
a preference for a particular brand.

Woody Allen	Olympia SM3
Agatha Christie	Remington Home Portable
F. Scott Fitzgerald	Underwood
Jack Kerouac	Underwood Portable
George Orwell	Remington Home Portable
David Sedaris	IBM Selectric II
John Steinbeck	Hermes Baby
Robert Louis Stevenson	Hammond
Hunter S. Thompson	Red IBM Selectric
Mark Twain	Sholes & Glidden Treadle Model
John Updike	Olivetti MP1 Portable Typewriter
P.G. Wodehouse	Monarch
Richard Wright	Royal Arrow

ON A ROLL

If ever any literary movement sought to sanctify the process of typewriting, it was the Beat Generation. "The typewriter is holy," intoned beat poet Allen Ginsberg in his poem "Footnote to Howl." Jack Kerouac, who favored the Underwood Portable, exalted the typewriter as his mechanical muse. By the time that he wrote his best-known works—On the Road, The Subterraneans, The Dharma Bums—even feeding paper into his typewriter had become an unwanted distraction slowing the stream of consciousness issuing from his Benzedrine-fueled brain. He took to typing onto a continuous scroll that, in the case of On the Road, ended up 120 feet (37 m) long. Only in the editing process was this exercise in free improvisation given some sort of order, or at the very least paragraphs.

Jack Kerouac (1922–1969) working on Trip-Trap: Haiku On the Road, a collaborative work with Albert Saijo and Lew Welch, in New York City, 1959.

"I'm going to get me a roll of shelf-paper, feed it into the typewriter, and just write it down as fast as I can, exactly like it happened, all in a rush, the hell with these phony architectures and worry about it later."
-- Jack Kerouac

Underwood Portable

WRITER'S BLOCK

The ten scariest
words in movie
history were
typed by Jack
Nicholson, who
plays Jack
Torrance, a
writer turned
axe murderer,
in Stanley
Kubrick's <u>The</u>
<u>Shining</u> (1980).
Shelley Duval
(left) plays
Jack's wife.

A page that might
have come from
Jack Torrance's
manuscript.

All work and no play makes Jack a dull boy.
   All work and no play makes Jack a dull boy.

All
work and
no play makes Jack
a dull boy. All work and
no play makes Jack a dull boy.
All work and no play makes Jack
a dull boy. All work
and no play makes
Jack a dull
boy.

All
work and
no play makes Jack
a dull boy. All work and
no play makes Jack a dull boy.
All work and no play makes Jack
a dull boy. All work
and no play makes
Jack a dull
boy.

All work and no play makes Jack a dull boy.
      All work and no play makes Jack a dull boy.
All work and no play makes Jack a dull boy.
   All work and no play makes Jack a dull boy.
All work and no play makes Jack a dull boy.
      Alo work and no play makes Jack a dull boy.

"Typing then was a muscular activity. You could ache after it. If you were not familiar with these vast keyboards, your hand wandered over them like a child lost in a wood."
-- J.B. Priestley (describing his first typewriter, equipped with a double keyboard for capitals and lowercase letters)

"I think the computer user does their thinking on the screen, and the noncomputer user is compelled ... to do a lot more thinking in the head."
-- Novelist Will Self (on reverting to using a manual typewriter in place of a computer)

"The biggest obstacle to professional writing is the necessity for changing a typewriter ribbon." -- Robert Benchley

"I pecked my stories out two-fingered on a Remington portable typewriter my mother had bought me. I had begged for it when I was ten." -- Octavia E. Butler

"I know so little about the typewriter, I once bought a new one because I couldn't change the ribbon on the one I had." -- Dorothy Parker

"I could not do without ... a typewriter, a supply of yellow second sheets and the time to put them to good use." -- John O'Hara

"This bunkum and stinkum of college creative writing courses! The academics don't know that the only thing you can do for someone who wants to write is to buy him a typewriter." -- James M. Cain

ON CELLULOID

Typewriters in the movies have
told stories, turned plots, evoked
sex and glamor, and inspired some
classic comic scenes.

>> 1. READY, WILLING AND ABLE (1937)
This musical ends with the hero
(Ross Alexander) writing a letter
to his sweetheart (Ruby Keeler). His
secretary helps him find synonyms
for "marvelous" to the syncopated
sounds of a typing pool. In the last
scene the lovers tap-dance across a
giant typewriter (shown right).

>> 2. THE BIG STORE (1941)
In this Marx Brothers movie, Harpo
takes typing to a new level, sowing
confusion at the keyboard.

>> 3. WHO'S MINDING THE STORE? (1963)
Jerry Lewis mimes typing in midair to a score by Leroy
Anderson in this comedy directed by Frank Tashlin. In
the orchestral version of the movie, a percussionist
"plays" the typewriter.

>> 4. ABSOLUTE BEGINNERS (1986)
One scene in Julian Temple's rock musical about teen
culture in 1950s' London features David Bowie dancing
across a giant typewriter.

>> 5. NAKED LUNCH (1991)
In this movie inspired by William Borroughs' novel of
the same name, a drug-fueled writer and part-time bug
exterminator played by Peter Weller imagines that his
typewriter has morphed into a giant insect.

>> 6. POPULAIRE (2012)
This French comedy directed by Régis Roinsard tells
the fictional story of Rose Pamphyle, who trains to
become the world's fastest typist.

42 KEYS TO VICTORY

During World War II, the U.S. armed services needed
600,000 typewriters. Buying them new was not an option,
because typewriter factories had been turning their
attention to producing war materials from November
1942—-Woodstock was the only exception. Instead, the
government asked the nation's typewriter owners to
sell any machines that were of a standard size and no
earlier than 1935 to Uncle Sam.

WHICH TYPEWRITER MANUFACTURERS MADE WHAT IN THE WAR:
Underwood   M1 Carbines
Royal   Machine guns, bullets, propellers,
                 aircraft engine spare parts
Smith Corona   M1903A3 rifles
Remington Rand   M1911A1 semiautomatic pistols

* * * * * * * * * * *
Hollywood helped
the war effort
by publicizing
the need for
typewriters.
Starlets such
as Maria Montez,
Gloria Jean, Susan
Hayward (left),
and Maureen O'Hara
were photographed
handing over hefty
machines to hunky
servicemen.

* * * * * * * * * * *

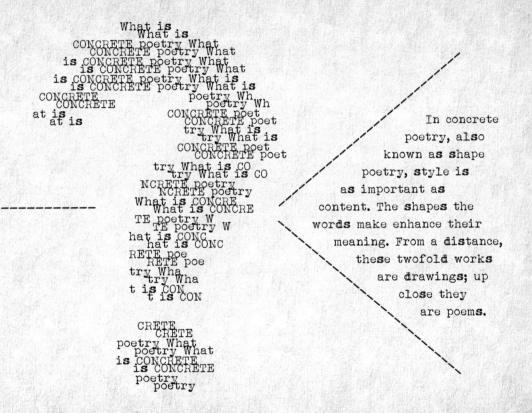

What is
What is
CONCRETE poetry What
CONCRETE poetry What
is CONCRETE poetry What
is CONCRETE poetry What
is CONCRETE poetry What is
is CONCRETE poetry What is
CONCRETE   poetry Wh
CONCRETE   poetry Wh
at is   CONCRETE poet
at is   CONCRETE poet
try What is
try What is
CONCRETE poet
CONCRETE poet
try What is CO
try What is CO
NCRETE poetry
NCRETE poetry
What is CONCRE
What is CONCRE
TE poetry W
TE poetry W
hat is CONC
hat is CONC
RETE poe
RETE poe
try Wha
try Wha
t is CON
t is CON

CRETE
CRETE
poetry What
poetry What
is CONCRETE
is CONCRETE
poetry
poetry

In concrete
poetry, also
known as shape
poetry, style is
as important as
content. The shapes the
words make enhance their
meaning. From a distance,
these twofold works
are drawings; up
close they
are poems.

BEFORE THE AVANT-GARDE

Concrete poetry didn't begin in the 1950s. Greek poets sometimes shaped their poems; later, in Lewis Carroll's <u>Alice in Wonderland</u>, the "Mouse's Tale" takes the shape of a tail. But the concrete poem came into its own with the typewriter, combined with an avant-garde sensibility.

THE DOM IS THE BOMB!

Dom Sylvester Houédard, a.k.a. the Dom (1924-1992), was a Benedictine monk and pioneer of concrete poetry. When not at his Olivetti Lettera 22, the Dom was either lost in prayer in a rural English abbey, or in London, collaborating with cutting-edge artists such as Yoko Ono. Seek out <u>Notes from the Cosmic Typewriter</u> to read more about the Dom's work.

MAKE YOUR OWN CONCRETE POETRY

>> Write or choose a poem or haiku.

>> Think of a shape that goes with it. Emily Dickinson's "'Hope' is the thing with feathers," for example, could be represented as a bird.

>> With a pencil, lightly outline the shape onto your typing paper. Decide where to put the words.

>> Type the words onto the drawing. You can space out or overlap the letters, or use symbols to create shapes or pauses.

>> Erase your pencil markings and admire your work.

Finalists of
Remington's Miss
International
Secretary
competition, New
York City, 1958

CHAPTER 4 :
* THE AGE OF THE TYPIST *

From automobiles to moon rockets, the 20th century was
the age of speed and the typewriter was no exception.
After Frank Edward McGurrin invented touch-typing
in 1887, typists were able to set down between 40 and
70 words a minute, in place of the 20 or so words most
people could achieve with handwriting or 30 words
per minute typing with two fingers. Encouraged by the
typewriter manufacturers, speed-typing demons vied
with one another to break records, with U.S. touch-
typist Carole Bechen tapping out a startling 176 words
per minute over five minutes in 1959.

"I feel that I have done something for the women who have always had to work so hard."
-- Christopher Latham Sholes, typewriter pioneer

Members of the League of Women Voters pioneer social media in preparation for the U.S. presidential election of 1920.

WILL TYPE FOR CASH

In the 19th century, nice girls who needed to work
didn't have much choice--governess, paid companion,
shop assistant. With the typewriter came a new sphere
of work for women: the office. The Central Branch of
the YWCA in New York City began offering typing
classes for young gentlewomen in 1881. When the Y's
grads landed the best jobs, other gals got the typing
bug. The money couldn't be beat: Proficient typists
earned up to ₡15 a week compared to a mere ₡6 for
working in a dry-goods store.

     By the end of the decade, nine out of ten new
employees in secretarial positions were women, and a
magazine noted approvingly that these "competent young
ladies" were "bringing into our business offices,
lawyers' offices, editorial sanctums, etc., an element
of decency, purity, and method." Spittoons and swearing
were on the way out, silk stockings and shirtwaists
were all the rage. But there was a downside to the new
job opportunities: Employers preferred to hire women
because they could pay them one-third less than men.*

A Bar-Lock machine

* In 2014, the
U.S. Bureau of
Labor Statistics
calculated that
women make
77 cents for
every dollar a
man earns.

THE NEED FOR SPEED

Typewriter manufacturers quickly
realized that speed mattered and set up
their own schools to train operators.
Other business people, spotting a gap in
the market, opened exclusive secretarial
schools for middle class girls filling
time before they got married. The crème
de la crème of these was the Katharine
Gibbs schools of Providence, Rhode
Island, New York, and Boston. As well as
repeatedly typing out "the quick brown
fox jumped over the lazy dog" (a phrase
widely used to build speed because it
includes all the letters of the alphabet),
the Gibbs girls took shorthand wearing
long white gloves and were drilled in
grammar. Teachers expected perfection and
students could get themselves expelled
for less. Using an eraser or correction
fluid was a cardinal offense.

TOUCH-TYPING
Frank Edward McGurrin, a
court stenographer from
Salt Lake City, Utah,
invented touch-typing
in 1887. Using a Sholes
& Glidden with a QWERTY
keyboard, he taught himself
to type without looking
at the keys. He toured
the western states giving
typing demonstrations.

Thousands of
typing schools
were set up
in the 1920s,
though not
usually in
the open air
like this school
in Germany.

In 1946, Margaret Hamma broke the record for typing on an electric typewriter (an IBM), by reaching an average 149 words per minute over one hour. The record still stands.

>> In 1926, Stella Willans managed 264 words of memorized text in a single minute.

>> The World Championship rules deducted 10 words from the total for every spelling error.

>> Speed-typing competitors wore beak-shaped visors with cloth extensions to minimize distractions and cut out the glare from lighting.

>> Underwood set up a specialized Speed Training Group to hunt out natural talent and train typing champions.

>> In a speed-typing competition of 1888, Remington offered a $500 prize for the team of typists that could most rapidly and reliably copy the 1,300-word Declaration of Independence.

Remington sponsored World Champion typist Rose Fritz from New York City to publicize its brand. Reporting on a contest in 1911, the Montreal Gazette described how she produced a "sonata of solid sound."

Typing pools were
like factories; a
far cry from this
1937 "typing pool."

THE TYPING POOL

Demand for typists was driven by the spread of
office work, which led to an exponential growth
in bureaucracy. Letters had to be typed out and
documents had to be copied, often not just in
duplicate but in multiples of 10 or 12.

      Bosses turned to the typing pool to get the
job done. Rows and rows of typists, overseen by an
eagle-eyed supervisor, churned out reams of paper
in a cacophony of clacking. Typing pools brought the
anonymity and boredom of production-line work to
the white-collar environment.*

* In 1938, one
insurance company,
Equitable Life,
reckoned its
accumulated
paperwork would
have risen to the
height of not one,
but 51 Empire State
Buildings (74,112
feet/22,589 metres)
if placed in a
single stack.

LIQUID GOLD
Before the advent of
correction fluid, typists
relied on unsatisfactory
erasers to correct
mistakes. Enter single
mom Bette Nesmith Graham,
from Texas, who invented
Mistake Out, a paper paint.
For years, she sold the
stuff from home, making it
in her kitchen blender. In
1956, she patented it under
the trademark Liquid Paper.
Bette sold the company for
$47.5 million in 1979.

* * * * * * * * * * * * * * *

TAKE A LETTER, MISS JONES

In retrospect, the Mad Men era of the
1960s seems like a typist's dream. Demand
for typists exceeded supply, and newly
qualified workers rarely had to wait to find
employment. But not everything was rosy
for the U.S.'s 3.5 million typists: They were
trapped beneath a glass ceiling and expected
to smile, even if through gritted teeth, at
sexist comments. Although some rose through
the ranks to nonsecretarial roles, the best
most women could hope for was to become a PA
to an executive, with a typist of their own.

MAD MEN: The glamour of early
1960s' hair and clothes, the
taste of a very dry martini
before lunch, the thrill of a
smoking—hot boss——if only all
bosses looked like Don Draper.

THE MOVIE VERSION

For Hollywood, the office had become a battleground, and the seat of power was not the typist's chair. However, in two iconic comedies of the 1980s, 9 to 5 (1980) and Working Girl (1988), ingenuity triumphs and the secretaries come out on top. In 9 to 5, an all-star trio of Jane Fonda, Lily Tomlin, and Dolly Parton kidnap their male chauvinist boss, then take advantage of his absence from the office to introduce job-sharing, a day-care center, and flexible hours. In Working Girl, a smart working-class secretary played by Melanie Griffiths trumps her idea-pinching boss (Sigourney Weaver) by stealing her boyfriend (Harrison Ford) and landing the dream job.

* * * * * * * * * * * * * * * * *

POWERING UP

Almost all typing was finger-powered until IBM made the machine that broke the mold: the 1961 Selectric. In the first year of production, IBM had orders for 80,000 machines; by 1986, when the machine was finally retired, more than 13 million had been sold. The Selectric featured a "golf-ball" type element that traveled across the paper, rotating and tilting to the correct letter as it went. This meant that there was no need for a moving carriage. Another innovation was changeable typefaces (remove the golf-ball typehead, replace it with one that had a different typeface, and voila!). Although not portable, the IBM Selectric, weighing in at just under 31 pounds (14 kg), was at least liftable. In 1973, IBM incorporated a correction ribbon that made repairing mistakes much easier. Typists all over the world cheered.

Although Hunter S. Thompson liked to take potshots at his red Selectric, novelist Jane Smiley felt more fondly about hers, calling it, "A writer's machine if ever there was one."

DREAM TEAM
- Horace "Bud" Beattie designed the golf ball element.
- Engineer John Hickerson worked out how to make the typehead spherical.
- Designer Eliot Noyes created the machine's elegant curves.

According to an IBM ad, the company's designers created the Selectric by "forgetting the past fifty years of typewriter design."

PRE—SELECTRIC
After buying
the Electromatic
Typewriter, Inc.
in 1933, IBM set
about improving
the product. In
1944, its Executive
model introduced
the proportional
spacing principle,
giving each letter
a different space
from narrow 'i' to
wide 'm.' By 1958,
IBM had sold one
million electric
typewriters, but
it was the 1961
Selectric that sent
sales soaring.

* * * * * * * * * * *

Public scribes tout
their services in
Istanbul, 1959.

THE TYPEWRITER WALLAH

For centuries scribes read and wrote letters for
the illiterate. Come the age of the typewriter, and
they typed them instead, often al fresco in public
places. In addition to letters, they composed
well-presented, correctly spelled documents,
for anyone who needed to communicate with
officialdom. Today, professional typists survive
on city streets from Rwanda to India to Myanmar.
But as literacy and computer use spread, the
typewriter wallah is fast becoming endangered.

Italian
manufacturer
Invicta shows that
even a supersized
typewriter can
be portable.

CHAPTER 5 :
* HAVE TYPEWRITER, WILL TRAVEL *

The first successful portable was made by
Blickensderfer way back in 1895. In an innovation
that seems very modern, the operator could change
the typeface, and even the language, by switching
to a different cylinder--some say it inspired the
IBM Selectric 70 years later. Despite the name, most
portables stayed on their owners' desks. For some
professionals, however, the word "portable" meant
exactly what it said. A portable typewriter was
an essential part of a journalist's toolbox: War
correspondents literally took their trusty machines
into battle.

THE FIRST PORTABLES

>> BLICKENSDERFER NO. 5 MODEL, "FEATHERWEIGHT," 1895
   - early models used a DHIATENSOR in place of the
     more popular QWERTY keyboard
   - A later version was made of aluminum (only
     recently developed) to reduce the weight

>> STANDARD FOLDING TYPEWRITER, 1908
   - a hinged carriage that folded forward over the
     three-row keyboard allowed the machine to fit
     into a small plywood case
   - later marketed under the L.C. Smith and Corona
     brand, the model remained in production until 1941

ONE LETTER AT A TIME
Index typewriters had
a dial that typists
turned to reach the
letter they wanted.
Although light,
they were extremely
slow to use and they
eventually lost out
to keyboard models.

>> REMINGTON PORTABLE, 1920
   - the first portable with four rows of keys
   - a lever at the side folded the typebars down when
     the machine was not in use, reducing the height
     of the machine to just 3 inches (75 mm)

A reporter **sets** to work on his Remington portable in a BMW Convertible, Paris, 1934.

TYPING IN COLOR

For many years, folks could have their typewriter
in any color so long as it was black. But with new
innovations in paint technology, typewriters soon
appeared in a rainbow of colors so that discerning
customers could match their machines to their decor.

Corona produced a range of
colorful typewriters, including
this one in Channel Blue.
The hard-wearing, spray-on
lacquered paint finish, known
as Duco, was developed by
DuPont in 1923 for use on cars.

This two-tone green Remington Noiseless Portable is from the 1930s. Although not actually noiseless, this typewriter is quieter than most since the typebar is stopped from hitting the platen with full force.

Made in the early 1950s, the Groma Gromina's bright red keys stand out against the dull purple body. This East German typewriter is one of the flattest typewriters ever made. Another ultraflat machine is the Rooy, made in France at about the same time.

OLIVETTI VALENTINE PORTABLE

Designed by Ettore Sottsass in 1969, this portable is so cool that it has a place in the permanent collection of New York's Museum of Modern Art. The chic design and lipstick-red color broke with traditional typewriter design.

>> Designer Ettore Sottsass called
his pop-art inspired icon "the
anti-machine machine."

>> Red is the most popular color, but it
also comes in white, blue, and green.

>> Orange spools provide a contrast
to the red body and black keys.

>> The plastic casing meant that this
machine was ultralight.

>> Olivetti still sells manual
typewriters, albeit made in a
factory in China.

Journalists at the 1962 Oscars where Patty Duke, as Helen Keller, won Best Supporting Actress.

MIGHTIER THAN THE SWORD

In the last scene of the movie <u>All the President's Men</u> (1976), Robert Redford (playing Bob Woodward) and Dustin Hoffman (as Carl Bernstein) sit in the office of the <u>Washington Post</u> typing out the story that will unleash the Watergate scandal and bring down Richard Nixon. Gradually, in a rousing cheer for the power of the press, the noise of the typewriter keys and the clatter of a teleprinter drown out the sounds of Nixon's swearing—in ceremony playing on a nearby TV.

Newsrooms are rather quieter now, but old movies such as Howard Hawks's <u>His Girl Friday</u> (1940) and the Tracy—Hepburn collaboration <u>Woman of the Year</u> (1942), as well as the TV series <u>Lou Grant,</u> keep alive the memory of the pre—1980s' newspaper office filled with journalists bashing out stories on typewriters.

* * * * * * * * * * *
CRAZY FOR CLATTER
Nostalgic for the constant racket of the typewriter? In 2014, the London <u>Times</u> broadcast recordings of old—fashioned typing through its offices in the hope of encouraging journalists to meet their deadlines. At about the same time, actor Tom Hanks launched the Hanx Writer app, enabling users to recreate typewriter noises on digital devices.
* * * * * * * * * * *

THE TYPEWRITER GOES TO WAR

In World War I, most typewriter
manufacturers turned to arms
production but one ingenious
little Virotype soldiered on.
Produced in France, the Virotype
was a small index device that
could hold a tiny piece of paper.
It was worn on the wrist like
a wristwatch for use in the
trenches or on horseback.

This pocket typewriter was produced
in watch form, and was made by
Torrani and Co. Sentences were
printed on the strip of paper, which
could then be stuck onto a page.

With her trusty typewriter
on her knees, Marguerite
Higgins buckles her helmet
in Nazi-held territory,
Germany, 1945.

WRITING FROM THE FRONT LINE
Marguerite Higgins (1920-1966)
arrived in Germany in 1945. She
and another reporter commandeered
a jeep, went behind enemy lines,
and were the first Americans
to reach Dachau concentration
camp. She took 22 S.S. guards
prisoner and later got an army
campaign ribbon. Higgins went on
to cover the Korean War and the
Vietnam War. During the Korean
War, General Douglas MacArthur
telegraphed: "Ban on women
correspondents in Korea lifted.
Marguerite Higgins held in
highest professional esteem
by everyone."

Reporters on the roof of
St. Peter's in Rome await
the white smoke signifying
the election of Pope John
XXIII in 1958.

E X T R E M E TYPEWRITING

Whenever, wherever, whatever, the need to type can be
extreme. A mountain in a blizzard, a desert island,
an iceberg—no place is too inhospitable for the
communion of mind and machine.

>> Scaling Mt. Everest in 1922, General Bruce, the leader
   of the expedition, said, "The Remington Portable is
   continually in use."

>> A typewriter accompanied Captain Scott's fatal
   expedition to the South Pole in 1911—12. The
   explorers used it to produce <u>The South Pole Times</u>.

>> In 1910, a Monarch typewriter was used in an
   underwater coffer to record the raising of the U.S.S
   <u>Maine</u>, sunk in 1898 during the Spanish American War.

<u>ALWAYS PACK A
TYPEWRITER</u>
Norah Gridley, cousin
of Abraham Lincoln,
and May Coleman (typist)
set up office (<u>c.</u> 1891)
outside Lincoln Log
Cabin, a family retreat
in Illinois.

THE TYPE-IN

On December 18, 2010, a new social event was born--the Type-IN. Organized
by Philly bikeshop owner and typewriter enthusiast Mike McGettigan,
the city's <u>Daily News</u> noted: "the Type-IN will give attendees a place to
type letters on a bar top--McGettigan is providing the stationery--trade
typewriters, and participate in a typing challenge."
    From the dozen people at that first meeting, the idea quickly spread
among the typosphere. Whether it's just three or four strangers in a bar
banging away at their laptaps or a full-scale convention, typers turn to
http://type-in.org, the main mouthpiece for the trend, to find out about
events all over the world, from Boston to Brisbane.

<u>ORGANIZE YOUR OWN TYPE-IN</u>
1. Find a bar or bookshop with a tolerant owner and
a big table.
2. Put up posters, send out fliers.
3. Bring at least three manual typewriters with you
for those who don't know it's B.Y.O.
4. Create your own typing competition.
5. On the big day, introduce attendees, hand out paper,
and get everyone typing.
6. End with a typewriter swap 'n' buy.

A young typist hits the keys at a Philly Type-In.

"Casting an eye over the cast iron casings of the classic typers scattered around my home, I wondered--who else was doing the same? I thought how grim the usual Laptopistan cafe scene is, where the brightest bulb in the room is the Apple logo, and considered what a gang of typewriters would sound like--instead of that chitinous clicking completely lacking in bells. A lovely pub manager agreed to share that hallucination and presto--a Type-IN."
-- Mike McGettigan, Philly typewriter enthusiast

>> <u>GLOSSARY</u>
Laptap (n)
Alternative name for a portable typewriter.

>> Laptopistan (n)
A place where laptops rule, e.g. Starbucks.

>> Type-IN (n)
A jam session for manual typewriters and the people who love them.

>> Typosphere (n)
The typewriter community, so called by aficionados.

INDEX

CREDITS